Book of Ideas

By Shubham Srivastava

Preface

Imaginations are the seeds of Innovation and Invention. This book contains well-documented ideas, theories, and models that could prove to be fruitful for changing the future. Every good idea needs a great investment.

Written with Hope,

Shubham Srivastava

ACKNOWLEDGMENT

I LIKE TO THANK EVERYONE WHO HAS GIVEN ME THE CHANCE TO LEAD, BE LED BY, OR OBSERVE THEIR LEADERSHIP FROM AFAR FOR BEING THE INSPIRATION FOR MY WORK. IT IS AS COMPLICATED AS IT SOUNDS TO TAKE AN IDEA AND TURN IT INTO A REALITY. INTERNALLY, THE EXPERIENCE IS BOTH DEMANDING AND SATISFYING.

1. Working Model for the establishment of Solar Panels on the roof top of gasoline to provide fuel for electric cars.

Abstract- The idea of solar power panel seeds up with the need to have electric-fuel stations for electronic cars. But Making up separate stations of electronic cars would not only take extra space of land but also takes an extra amount of infrastructure establishment and labour to establish. We could upgrade conventional gasoline stations to modern solar power panel stations by adding minor adjustments to them with the help of this working model.

Index Terms- Electric Cars, Solar panel, Electric cars charging station, Power Station, Charging Station

I. INTRODUCTION

The working model for the installation of solar panels on the rooftop of gasoline stations to extract clean solar energy and convert it into electric energy with the help of an inverter. The inverter sends down the electrical energy to the charging station.

II. IDENTIFY, RESEARCH AND COLLECT IDEA

In the recent trends, electric cars\transports are viewed as an alternative to oil-petroleum-based transportation. But one of the main barriers in buying cars is the lack of charging power stations that work as fuel to these cars. Thus, consumers are less interested in buying them.

Many people came up with an idea that an electric power station needs to be established for these cars. The electricity in most of the gasoline is harnessed by solar energy. The same energy could be used for the power charging station.

But, setting up a separate electric charging station as fuel for these cars is time-consuming and it will provide fewer options for these cars to fill up.

III. STUDIES AND FINDINGS

With little modification on petrol-CNG pump could serve as a station for electric cars:

A. Solution:

- Setting up solar panels on the roof-top of the station/petrol-pump to generate electricity.
- The solar cells will generate power which in turn will store in the batteries. These batteries will transfer energy to inverters.
- The inverters will send the electrical energy to the charging station through wires. The charging station will have arrangements of apparatus to charge the vehicle.

B. Units.

- The setup will have a charger like EV-T2G3C-1AC32A-5.0M6.0ESBK01 with maximum capacity of Maximum Voltage Rating (VAC) 250VAC/30VAC.
- The station will use a 65KW Superfast charging model that could charge up to 80% in half an hour.

IV. CONCLUSION

The working model is the solution that could encourage people to shift from petrol/diesel transport to electric transport. The ease of availability of electric stations on the same gasoline site will also save the lands for other purposes.

V. REFERENCES

[1] Electronics & Communication Engineering https://nptel.ac.in/courses/117/102/117102059/

Principle Working of Solar-panel generating electricity for electric cars.

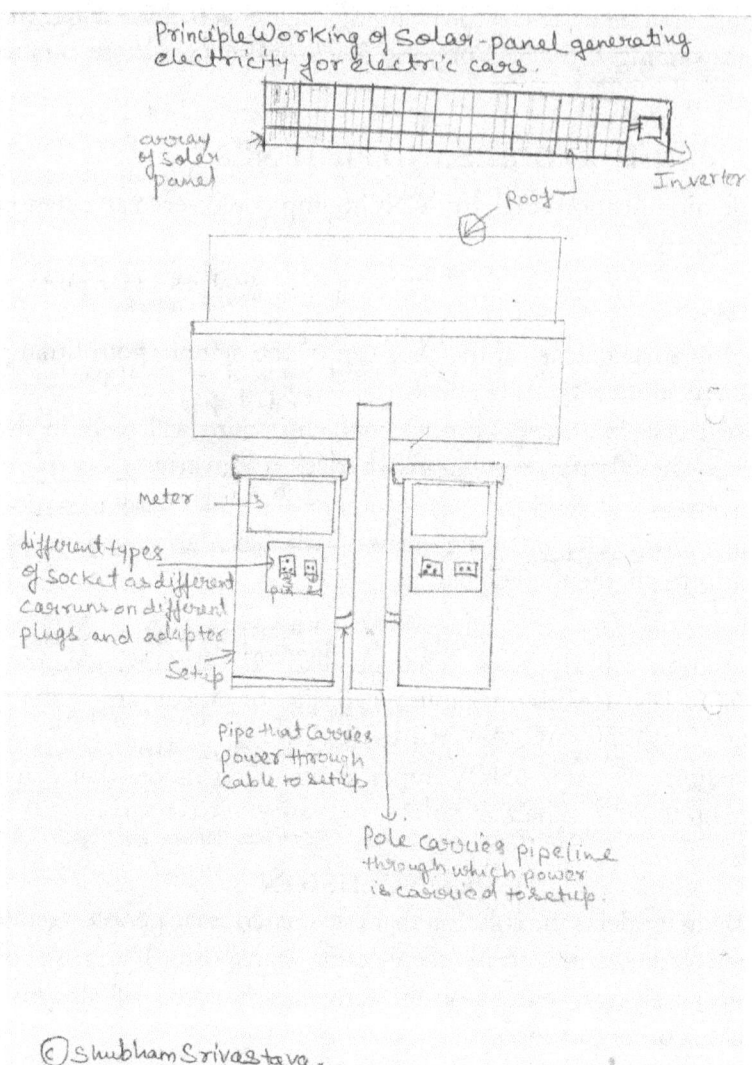

array of solar panel

Inverter

Roof

meter

different types of socket as different car runs on different plugs and adapter

Setup

Pipe that carries power through cable to setup

Pole carries pipeline through which power is carried to setup.

© Shubham Srivastava.

2. Medmax, an all-encompassing health care app to bring the hospital to your doorstep.

Abstract- The idea of Medmax seeds up with the intent to provide hassle-free speedy delivery of health-care. Often, critical patients lose their life due to improper medical care. Medmax, is all you need.

Index Terms- Medmax, Health-care app, Hospital, Medicine

I. INTRODUCTION

MEDMAX IS RELATED TO HOSPITALS IN THE SAME WAY TRIVAGO APP IS RELATED TO HOTELS. YOU COULD SEE ALL THE NEARBY HOSPITALS IN YOUR LOCALITY WITH THE TYPE OF MEDICATION YOU WANT. THE AVAILABILITY TO DOCTORS, THEIR TIMINGS, APPOINTMENT FACILITIES, BOOKING OF EMERGENCY ICU CASES THROUGH THE APP WITH FULL-PROOF AVAILABILITY OF THE NUMBER OF REAL-TIME VACANT ICU WARDS IN THE HOSPITAL. WITH THE HELP OF THE AFOREMENTIONED ATTRIBUTES, YOU CAN ALSO BOOK AN AMBULANCE THROUGH THE APP FOR THE PATIENT AND TIME TAKEN BY AMBULANCE TO REACH HOSPITAL. THE APP ALSO SHOWS THE TOTAL CREDIT REQUIRED FOR THE DIAGNOSIS. IT SHOWS DISCOUNTS GIVEN BY THE HOSPITALS ON TREATMENT AND MEDICINES.

II. STUDIES AND FINDINGS

The first need for urgent a cost-efficient medication facility arises from my own experiences. For Instance, look at this data, shown by the Ministry of Transport, presented by Govt. of India. The Ministry of Transport, Study has reported that 151,113 people were killed in 480,652 road accidents throughout India in 2019, an average of 414 a day or 17 an hour. The numbers were huge, half of life lost on the streets could have saved by timely delivery of treatment. Medmax is made for the same.

Another good case for urgent diagnosis is stroke, cardiac arrest, and other heart related disease. If patients get urgent treatment 45 mins after the occurrence of stoke. Thereby, increasing the chance of life manifold.

III. CONCLUSION

THE WORKING MODEL OF THE MEDMAX APP WILL SURELY SAVE PLENTY OF LIVES SOON. HOWEVER, IT NEEDS BIG SEVERS TO HOLD SUCH A FRAMEWORK.

3. Establish underground tanks near the river banks to store the rainwater when it rises in the monsoon.

Abstract- The idea of underground tanks near the river banks to store the rainwater when it rises in the monsoon to deal with water crisis at large.

I. INTRODUCTION

WHEN THE MONSOON ARRIVES, THE WATER LEVEL IN THE RIVER RISES AT AN ALARMING RATE; WHEREAS IN SUMMERS, THE SAME RIVER SUFFERS WATER SCARCITY. A LARGE AMOUNT OF WATER GOES INTO THE BAY OR OCEAN IN THE MONSOON, WHILE GROUNDWATER RESERVES KEPT DECREASING AT A RAPID RATE. IF THE SAME AMOUNT OF WATER SOMEHOW TRANSFERS FOR RECHARGING GROUNDWATER RESERVE COULD DO WONDERS.

II. STUDIES AND FINDINGS

The working model contains a series of underground tanks connected through pipes shown in the figure. These pipes will contain Nano filters that cleanse the impurities of water and have discharged tubes in the bottom of the tank to ejected the same water into the ground. A few cusecs of water will be sent to the irrigation and water ministry of the city for domestic purposes as well as avoiding overflow in tanks at the same time.

III.CONCLUSION

The working model has plenty of possibilities to save water in the near future. The Delhi model of tapping Yamuna water is quite similar to the model presented. It has already saved plenty of water last monsoon.

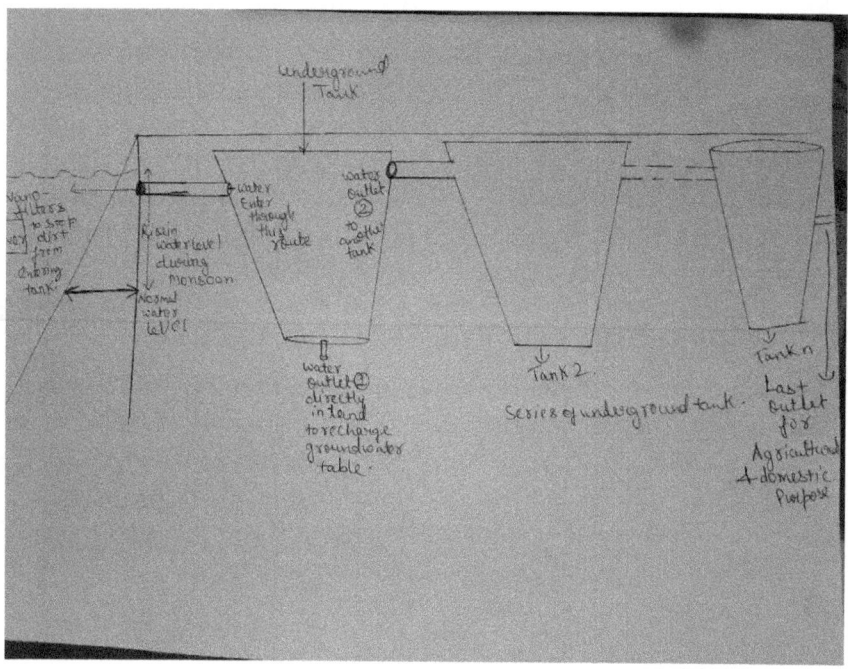

4. JOIN ALL THE RIVERS TO AVOID SEASONAL DROUGHTS.

Abstract- The idea of connecting large rivers are indeed visionary to reduce drought at large scale in the drought prone and tropical regions.

I. INTRODUCTION

THIS WORKING MODEL WAS FIRST INTRODUCED BY INDIA'S VISIONARY PRIME MINISTER LATE ATAL BIHARI VAJPAYEE. CONNECTING ALL THE RIVERS WAS ONE OF HIS VISIONARY PLANS TO TACKLE DROUGHT-PRONE AREAS AS INDIA USED TO DEPEND ON MONSOON, ESPECIALLY IN RURAL AREAS. THE CONDITIONS ARE LOT BETTER SINCE WATER-IRRIGATION FACILITIES ARE INTRODUCED, BUT THE ALTERNATIVE OF THIS MODEL IS STILL NOT AVAILABLE.

II. STUDIES AND FINDINGS

Monsoon is irregular and highly dependent on the precipitation induced by asymmetric heating of land and sea. The absurdity in the climate and seasonal winds are also some of the factors. This leads to drought in many areas as many water bodies such as rivers, lakes, and ponds fail to recharge in the monsoon.

To avoid this, all rivers are needed to be surrounded by forest and connected with other rivers. Although, it seems to be impossible to connect all the rivers in India. But, two adjacent rivers can be connected which could curb the drought to some extent.

III. CONCLUSION

The idea of rivers is somewhat nasty at first place but could be worthy after implementation as it could save the lives of million.

5. Install sensors in parking spaces of multi-storey car parking system to give real-time information of spaces which are available for parking.

Abstract- The idea of sensors-equipped parking system with the intent to ensure hassle-free parking on the malls, hotels, hospitals and other crowded surfaces.

I. INTRODUCTION

Parking sensors system requires AI, Embedded Systems, Cameras, and computer programming. Each component plays its part to perform as the complete parking system as a whole. It will make parking seamless for customers and visitors.

II. STUDIES AND FINDINGS

The parking sensors will fit on the pole would be situated at all four corners of the parking area. There would be one micro-camera on one of the poles to ensure car safety from any mishap or burglary.

When the car arrives at the parking area, a valet would look into his monitor which displays the list of vacant parking spaces in the parking system. He would generate a token from the computer and give it to the visitors.

When the visitor successfully parks the vehicle in the parking region. The sensors will beep once and generates the token message 'Occupied' on the monitor.

III. CONCLUSION

The parking system will generate real-time vehicle information and ensures safety of the vehicle as well.

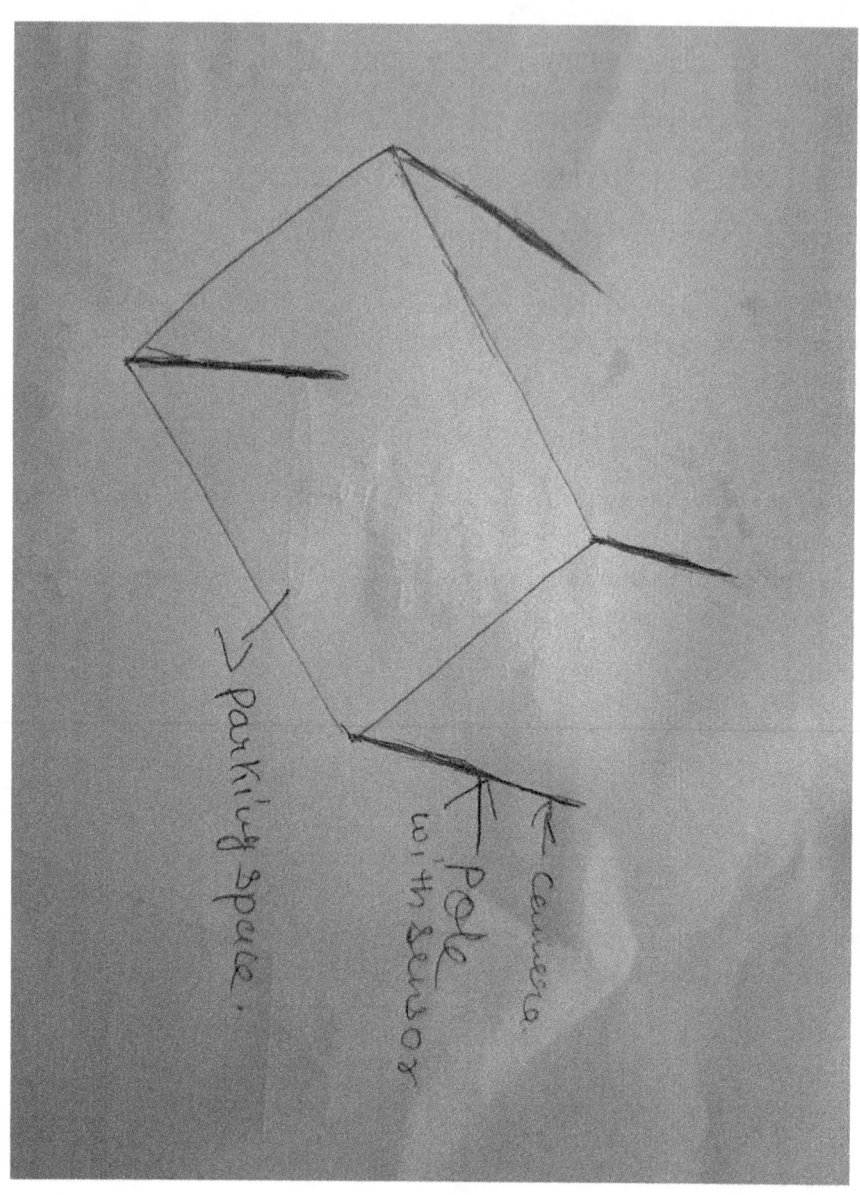

parking space.

camera

pole
with sensor

6. Design a thermos-sensor bath tub.

Abstract- The idea of thermos-sensor bath tub with the intent to ensure safety of children and adolescent from your app from excessive overheating of water.

I.INTRODUCTION

Thermos sensors would fit inside one of the edges of the bathtub to ensure temperature of the water. These thermos-sensors not only measure the regulate the temperature but also turns it off by generating message on the app.

II.STUDIES AND FINDINGS

The user will set up the temperature through the app. App turns on the geyser that is kept on until the water reaches the required temperature. Once the temperature is achieved, the geyser ultimately turns off. It generates the message of 'work completed'. The whole process is performed within 20min.

III.CONCLUSION

The robustness, efficiency, and safety are the points on which the appliances will work upon.

7. Arrangements of ranking dead websites from remove the world-wide-web garbage.

Abstract- Every year tons of content are added on website. Some of the website remained no-follow on the internet for the years. Such website has no visit whatsoever in a decade needs to be considered as non-Seo friendly and removed from the web.

I.INTRODUCTION

There are many kinds of browsers like Opera, Chrome, Wave, DuckDuckGo, Brave, Tor, and Vivaldi. These browsers have crawlers that index and rank the content based on keywords. Removing non-Seo-friendly websites that have not been visited even once in a decade could remove internet trash up to some extent.

II.RESEARCH AND FINDINGS

Recently, Google has announced to remove non-active google photos and google drive accounts after 5 years of inactivity. This will remove free-estate accounts, bots, inactive accounts from the world wide web. The same idea could be applied to websites that have been inactive for a decade. However, it could not remove a lot of data but ensures privacy threat, Identity Misuse, removal of bots for a credible platform, and many other phishing websites. Every day huge terabytes of data are stored on servers, applying this could remove a huge load of redundant data, non-worthy data and make WWW hassle-free.

III.CONCLUSION

Removal of dead websites to remove dead websites from the World Wide Web will induce sustainable policies in the virtual world.

8. Install a free product shop in every city where brands can showcase their new product.

Summary: These kinds of shops would introduce new brand products for free to a larger audience to record feedback for their video testimonials. Every positive testimonial will shoot brand reputation to the new set of customers. It could be worthy for new start-ups to create buzz among the audience instead of paying a hefty amount of money for ad-marketing in the first place.

9. Thought Camera by Nikola Tesla

Summary: There are plenty of weird Nikola Tesla inventions but the introduction of AC circuit generators defined new ways of energy transmission in the last century. Futurist and visionary to the core, there are plenty of research papers on his name that could become reality in near future. One of my favourite picks from Nikola Tesla is undoubtedly thought camera. Weird, but we are very close to achieving it. Computerized Tomography, Magnetic Resource Imaging (MRI), that performs neuroplasticity is the first step in this direction. The second step was introduced by the development of software and embedded system by Intel to provide a way of communication to Stephen Hawking, a renowned scientist whose ability was restricted by motor neuron disease. The machine converts his thoughts into voice with the help of AI. The third step could be considered as Bio-Hack trails by Elon Musk draws inspiration from Nikola Tesla. Bio-Hack is made with the intent not only to read human-mind but to rectify disabilities attached to the body to unlock the full potential of a brain. It could be possible near future.

10. Meat alternative.

Summary: Plant-based meat alternative is a thing now, many organizations such as Memphis Meat, Mosa Meat, Alph Meat, BlueNalu, Finless Foods are tirelessly working in the production of lab-grown meat. These meats are grown in a lab without harming any animal. The lab-grown meat has all nutrients of the original meat and is in the final phase. The only thing that lacks is the feel of bones in the meat as meats produced by the lab in the minced form. This will reduce abuse on animals and illicit animal farm practices that produce environmental pollution to the core.

11. Air Pollution and seasonal air-pollution

National Capital India suffered seasonal air pollution in winter. The air quality becomes dense that traps smog of the region. This problem persists in many other regions of the world. To tackle it, An IIT-Delhi alumni founded Kriya Labs that made solutions like Tableware, Art, and Craft Handbags, Moulded and Rigid Packing Solutions. These solutions not only reduced stubble-burning practices but also reduce the plastics and highly non-biodegradable products use to the core

12. Alternatives for plastic-packaging

Plastics are one demon that humanity want to get rid of it as soon as possible. It is poison for flora and fauna alike. Scientist Bio-engineer a bacterium name Ideonella Sakaiensis that consume plastic as an energy source. Culturing this bacterium to eat plastics could be real thing in near future. Also, uses of bio-degradable bioplastics like corn and sugarcane can totally remove plastic packaging from market. NatureFlex is a range of specialist packaging films produced by Futamura to provide packaging content choices that provide great support to environmentally friendly packaging needs for customers. It uses wood pulp as an alternative to cellophane for packaging.

13. Solar-equipped Society

There are plenty of opportunities in developing solar-equipped zero-emission smart cities. From infrastructure that entraps solar energy to the bicycles that suck pollution and turn it into oxygen.

References: https://www.youtube.com/watch?v=eVFYhbHpfqU
https://www.youtube.com/watch?v=AQuI_AhENg0

Honorable Mention.

Genetic Engineering of trees to increase oxygen production.

Some trees spread more oxygen than others. The growth of the population requires intelligent environment systems to cater to the needs of the masses efficiently. Genetic engineering is the process of using recombinant DNA (rDNA) technology to alter the genetic configuration of an organism. It is used to increase the desired trait/behavior of a crop/plant/tree or an animal. The same technique could be applied to plant/trees to increase the amount of oxygen they exhale for other lifeforms. Examples of such trees are Douglas-fir, Spruce, True-fir, Beech, Maple, Syzygium Cumini, Alstonia Scholaris, Murraya Koenigii, Terminalia Arjuna, Saraca asoca, Azadirachta Indica, Ficus benghalensis, and Ficus religiosa.

About Author

Shubham Srivastava is a software developer, blogger, and content writer who loves to listen to music, read books, and have some good food with friends, introvert to the core who loves to observe or analyze to get to the bottom of things. He has been publishing articles on a range of topics from economics to business, finance, politics, current affairs, international relations, etc.

Other Works

- Read 'Quotes for Life' available on Amazon, Google Book and Notionpress Store.
- Read 'Do gaj Zameen' (Hindi) available on Amazon, Google Book and Notionpress store.
- Read 'Two Yards of Land' English Version available on Amazon, Google Book and Notionpress store.
- Read 'Mafia' available on Amazon, Google Book and Notionpress store.
- Read 'Satangel' available on Amazon, Google Book and Notionpress store.
- Read 'Mahakaal' (Hindi) available on Amazon, Google Book, and Notionpress store.
- Read 'How to cheat death and improve life naturally' available on Amazon, Google Book, and Notionpress store.
- Read 'Awaken' available on Amazon, Google Book and Notionpress store.
- Read 'Emotion: An acrostic poetry cuisine' available on Amazon, Google Book and Notionpress store.
- Read 'Bio-Terror: you're being watched and monitored' available on Amazon, Google Book and Notionpress store.
- Read 'Queens of India' available on Amazon, Google Book and Notionpress store.